The Universal Code

Published by Nicholas Betar
Copyright 2014
2nd Edition 2016
3rd Edition 2023
Contact: c/o Australia post office,28-30 Bay Street,
Southport
Qld 4215, Australia

~~~

ISBN: 978-0-646-90371-2

Table of contents

~~~~

Preface

This book is a researched and attempt at finding the aspects to the whole event that is numerals and numbers. Constructing a reduction of the variance of the cube in correspondence of architecture and technology in application of letters and numbers.

Chapter 1. The Alphabet and numerical terminology

Would you believe that the sum total of alphabetical and numerical information can be present through this very algorithm below in *illustration 1*. The manual layout of this alphabetical/numerical algorithm has with it an exceptional amount of information. For instance, where the master numbers

2

converge on this algorithm, the numbers that make it happen are positive and negative in nature. They are the binary numbers zero and one. They can be looked at in the metaphysical sense as male, the one; and female, the zero, another way would be the phallic symbol, the one; and the circle of life, the zero. To this as translated through the binary numbers, the converging, meaning lining up, as above and below, we have result of this master number through six plus five, equalling eleven. It so happens to be sequenced three time across the algorithm on both sets, above and below on the number line, though not the alphabetical line. F,P and Z are part of the convergence of the master numbers.

Illustration 1.

1 2 3 4 5 6 7 8 9 0 1 2 3 4 5 6 7 8 9 0 1 2 3 4 5 6

a b c d e f g h i j k l m n o p q r s t u v w x y z

0 1 2 3 4 5 6 7 8 9 0 1 2 3 4 5 6 7 8 9 0 1 2 3 4 5

The break-down of the algorithm is started through the knowledge of numbers and letters used in a gridded format, based on the space around us. The space is commonly known in metaphysical knowledge as the 'ether'. The ether is translated to our planetary bodies for example, the Earth, the moon and other planetary bodies in the solar system. This becomes the history of mapping the planets of the solar system, for example Galileo who began this course of action with Telescopic devices using precise measurements in his instrumentation. We can of course go further back in history and recognise further developments in measuring the planetary bodies and this has been found through many cultures of this type of work. What this indicates is a map, a map that is since it's history of mapping the planetary bodies, gives us, the Zodiac, though before this is considered and is indeed part of this, the binary O and 1, is the bookend explanation to both the zodiac and the colour tone and frequency indwelling within those. It has also the base of those, the colour tone and frequency for a template to whatever structure as I outline further in my book with an explanation of architecture in relation to the letters and numbers. The Zodiac is made of twelve signs, each representing a part of the night sky, as pre-recorded through history, this brings the concluding factor into this.

Chapter 2. Eleven becomes One hundred and eleven

The conclusion of having reached eleven is to realize it's further migration to a higher number, even a vibratory note if a frequency is being measured. Zero, the circle of life gives the charge to the One or the phallic symbol, what resonates here, is Earth bound, where as what resonates in the 'ether' is space bound. As has been mention by Hermetic philosophy, as above, so below, that is the opposites.

In regards to opposites and equals, the quint-essential problem is determining when a number translates or for that matter transmutes, in the metaphysical it is more determinable, in theory or formula, the same can prevail, at this point it becomes mathematical and maybe complex, so the algorithm encompasses so far, numbers, letters, theses being the symbols, then the link to binary. The phallic symbol the number one, the male and the circle symbol, the circle of life the female onto then the planetary in how numbers and letters were used to indicate measurement of instrumentation as in astronomy and astrology with Galileo and with many cultures around the world measuring and receiving ideas from these numbers and letters. I will exemplify further what I'm reiterating here in the next chapter.

Chapter 3. The idea of 'being' of the numbers

The idea is of the being of numbers and letters made into a word of understanding. Our universe is vast, we cannot assume we are alone in it. Why do we assume sometimes, the vast majority of things that stop thought is a non committal and non communication level on things, a very human trait! But we can break this from time to time, we can experiment, this is human nature and we can have the work involved making us feel all the many traits and ideas but never concluding on an idea, the trait is more of a habit being either negative or positive or neither of the two.

The 'ether' holds the idea, we just have to recognize it and is a creative factor, in fact probably the most creative factor of traits because it's just a word, like an idea. The idea is a number, the number is twenty-six. I'm outlining the obvious, though there appears to be in older texts cultural or otherwise of further letters, the average today represents the truth of our languages, all sharing the average letters. Now, this determines the number on the algorithm that is exemplified to this. Twenty-six represents twenty-six letters, these twenty-six letters are representing the World's, these can be represented best as is heard from Hermes Trismegistos, "The world above and the world below." The world above and the world below have the representation of the Sky or Heavens, the world below is a representation of the Ground or Earth, whatever is closest to the ground is what we participate in, life, the world, what we use to participate with it, are letters and numbers, these are from the ground source. I have shown on the logarithm, the one is for-most representing the master numbers, being that of eleven twenty and lastly thirty three. Binary represents space or void. What makes for the ground source is the number one, or male for the source.

What makes for the Sky or Heavens, is the number that constitutes a factor of divisible though in the algorithm, the invisible, zero and zero only equals, the circle of life. It has to be made into a number twenty six. The reason is binary, the application of space. Twenty-six becomes two-hundred and sixty. Semantics on this idea can be applicable anywhere between twenty-six and two-hundred and sixty, to cut this short, you can now add a zero, a circle of life to each concluding sphere of space, much like an onion. This explains the grid of space and environmental factors and so the grid will contain other new logic, understandings and principles. Varying consciousnesses that I call 'illusory' being illusion operate in the dynamics of mankind's contributions to the physical realm, for example, the impact and imagery from television and the sound produced from a radio, any varying logarithmic patterns operate at subtle frequency giving these phenomenon and all operate at any given time in these realms. The extent of these phenomenon and at which and what goes on each generation, is the changing patterns within each grid learning and their structures of uniqueness within each structure, there is the question of the cube in relation to the grid.

Chapter 4. The cube

The grid has the three dimensional grid cube attached to each grid square, for where ever there is a square, there will be a cube. The cube and not unlike ourselves of that individual space that both the cube and ourselves may occupy, has other grid cubes that are either larger or smaller, yet no less significant, grid cubes of consciousness, are for the grid cube that we exist within at whatever level, height breadth and depth, they are always around us at any one time.

As in algebraic equations, the set group of numbers is apart from the primary of the group, example, $7 - 6 (+1) = 2$ yet if this is changed to $7 - 6 (-1) = 0$; What accounts for the difference is 2 and again this $7 - 6 (+-1) = 0.5$. Overall, the same goes for multiples of a number, for instance, a 0.5 or 1 added to any number in a set group or individual number; another example, we have $7 - 6 + (3 \times -1) = -2$, although change this to $7 - 6 + (---1) = 0$, just as the opposite $7 - 6 + (+++1) = 2$. If now the character is individualised and so bracketed as $7 - 6 + (-)+(-)+(-)1$ as individualised separate groups for -1, it is then -2. The same in positive is $7 - 6 + (+)+(+)+(+)1 = 2$ This demonstrates a - 1 or +1 are one and the same from the perspective of the cube and of space in performance with numbers and letters of the algorithm and the illustration below, the individual cube is both negative and positive, it is also as demonstrated with the algebra equations, indivisible to multiple performance of cube interaction. Energy manifestations would be naturally random and so a different reflection but the same grid.

Grids represent language and language represents our reflection of where we are represented in the space of language, the dimension of it. The world holds the only answer to our organic structure.

Chapter 5. Mapping through to architecture

Simplicity is the apparent determinant including that earliest of mapping of any project that is undertaken as for example, a boundary line on a field to the latest of satellite technology taking a picture from space. The same goes for general planning because history is integrative our whether we do something on a small or large scale. Through history, mankind has endeavoured and successfully undertaken these opportunities. Earlier on in history, theses have started with simple prerequisites, be it based on tools for a base to map and build a house and integrative features of practicality.

Here the number system has worked from earlier times using mathematics, geometry and set squares, right to the present day in delivery of functional modes of working , these however were not always as paramount as they are today in the use mapping and architecture using computer and technological processes of design and architecture. Programmes such as 'Cad' are in use for such requirements of making and designing, the same is related to the technology, for example aircraft, though, this being more recently.

Flight and technology is and has integrated throughout our history alongside the developments of mapping and architecture. The early pioneers of flight brought us into a recognised history of flight. In ancient history, the myths of Indian scripts of texts tell that the working of flight was prevalent back many thousands of years ago with a vehicle, the scripts tell us of their Vimana's, being saucer shaped. Because our history is mixed with accounts whether historically mythical or otherwise, we are never the less dependent on this to further development in areas of expertise of thought and practice.

Numbers have played a pivotal role in historical workings, this development has shown another feature of the number on the top line of the algorithm. As illustrated;

Illustration 2.

1 2 3 4 5 6 7 8 9 0 1 2 3 4 5 6 7 8 9 0 1 2 3 4 5 6

I've outlined just the top line to show the correspondence with our present history.

The history and future, if we could say are one and the same, then where would the catalogue of the Akashic records be housed and what would they look like and like architecture, what would the structure look like? For the formula of mathematics or the length and structure of written words.

This shows the break down description of what might be Earth. void and pressure=darkness and Air is also void with no pressure=light. Therefore for example the letter 'A' holds information, the same would go for the number '1', though what can be vice versa is the pressure or no pressure, for example, an algorithm generated piece of information code that is put into a computer is going to have more pressure with more numbers or letters or both then say with fewer or just one number or letter, this alternative description is to do with 'space' on a computers drive although this can be looked at as a type of technological microcosm of the greater space outside of that realm. So information that can be the same or one character or unit or units, again for example A is A as a unit through to Z or in algebraic terms, A is one through to zero as a group, depending on how you approach the usage of the unit or character, the former could also be multiplies of the letter or number as another example, algebra produces sub-sets with mixes of numbers and letters and this could be of an algorithm.

Taking into account the logarithm, the unique structure of it is prevalent to equal and odd numbering taking into account letters. It's usage in pyramid design and structure will have a zero and a one in the building of the sides of the structure. There is a triune usage being three points in architecture. In this matter, mankind has discovered knowledge, placing it into engineering in letter and number, to give the practical exemplification of universal structure and which is the triune, the three points in nature as the basis for structure being solid and symbolic.

Working from below the ground to the surface and then to air, there is interpretation to each of these areas. From the physical and the metaphysical point of view, below is pressure, it is dense under the ground from rock structure, it is mounting pressure the further you go down. The surface, our landscape is unlike the ground below, though similar pressure or air pervades here and air has different pressures, Earth, Air, Fire and Water make up the basis of the compositions of these, with different chemical and composition both to the rock structures and the air and gas pressures, with all of these having a gauged number and periodic letter, the algorithm we have technology utilizing Earth, the breakdown of materials into their components, then go into building structure, using construction materials which is for architecture, then mapping is the structure of the inner architecture through mind and intellect, the inner architecture description, location geographically and utilizing the algorithm number and letter as formerly described both in scaling and drawing the print plans, the third is air, it represents technologically, flight and the structure is aerodynamic and whilst it arrangements fits into aerodynamic shapes and materials, the structure is born.

The Quabbalistic term for the usage for name used for God has the expression of a zero and a one expressed. This is a good example of for instance in the Akashic records, a sub-conscious reasoning for what God is and we have structure for the records. To my mind, the structure, indeed the architecture of the Akashic record, our spiritual depository is none other than the zero and the one being a hole in the structure of space. This could be termed, a portal or an opening and almost a metaphysical meaning with a biological overtone, again as mention the zero is female, for 'vagina' and the one is 'penis', the knowledge could be termed as this. Whatever this transmutes into, could be just called knowledge and to go further into terms about this, is back to the present, the present represents Earth matter and out of minerals from getting the out of the Earth, mined, drawn up in the plans and made to be the completed product from Earth to Air and again drawing on the factor of Fire and Water in the displacement of the working materials, the rest is mapping it out.

Culture is predominantly elucidated from its history, many new and not so new inventions in the field of architecture, written workable plans for the design and manufacture of Earth, Air, Fire and Water in the product sense is conclusive engineering and found at ground level by this I mean that we have a formula which uses algorithms or hints at that to use the terminology and can be linked to Phi or (golden mean ratio) 0.1618 and into infinity from the numbers from there. When the circle is squared, six represents natures formation. as with *illustration* 3 of the golden mean below:

Illustration 3.

It can be used as a type of measure in architecture gauging it is neither one or zero, it could be all the numbers in connection with the golden mean, unwinding and winding because of transfer of information this on a biological scale. The nautilus shell demonstrates that to build the shell took this factor into account as the sea creature grew. Like the zero and like the one, it is neither negative or positive, they are interchangeable yet hold to a set of principles and laws. In theory, like a strand of DNA, that winds and is a portal, it is an opening to the Akashic awareness within.

Illustration of geometries of the zero and the one. What they constitute. How they function in terms of geographical and geometric design in nature.

Illustration given below of the set of separate but inter-related geometries to structure and sub structure on microcosmic and macrocosmic levels in relation to atom the fields of algorithmic inter-relation and interpretation. Assuming there is no wall around the structure that blends within this, the cell is never a cell without a structure that is understandable to human logic and thinking. There is a membrane that surrounds this and there is an energy field within the nucleus. They confer as to the following with *illustration 4..4A*. Shows the perpendicular lines of structural core. *4B*. Reveals as I've outlined, the inverse and obverse of the pyramid structure with pyramid tips touching each other or the point within the circle structure. *4C*. If the pyramid is drawn within the structure, then it shows the more adapted structure of the Tetrahedron. *4D* The structural core then has around it, the tetrahedron structure, which in its structure also shows the seal of Solomon. That which is below is also above. It demonstrates how the circle helps to manifest this above and below from a side view of the pyramids with tips touching the point.

Illustration 4....4A,4B,4C,4D

4A, 4B, 4C, 4D,

4E,4F are further exemplifying of the same showing different regions of use in geometry

4E

4F

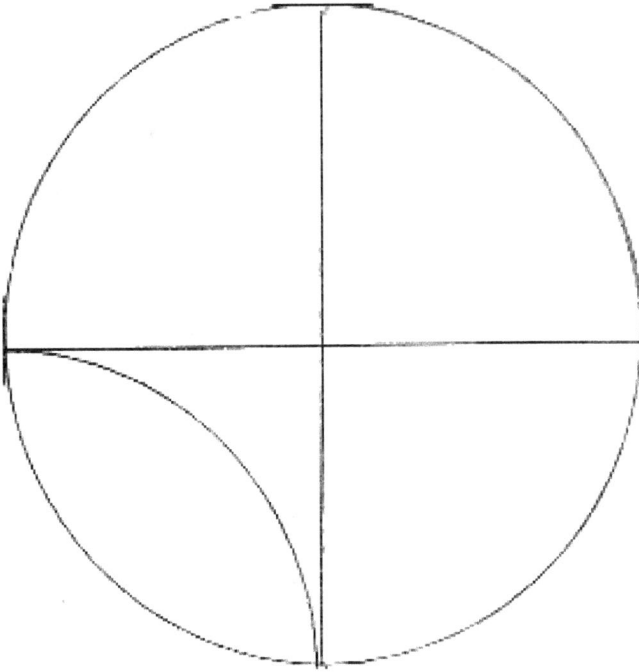

4G

This illustration is comparative of the flower integration with a semi circle added to the squaring of the circle in one quarter panel of the illustration of 4G. It is indicative of the continuance of structure both in nature and architecture, which of course parallels in mapping, geometry, (geology i.e.: geomancy) and technology, these include the application of them. The quarter image if turned on its point, pirouette, the image assumes the typical squaring of the circle image.

There is an interesting doubling effect with the number sequence that is prevalent in the numerical/alphabetical distribution of the algorithm. For instance for the number of 5's on the top line of the algorithm is 15, the 6's on the bottom, make 18. 1+5 is 6 and 1+8 is 9. This would have it symbolism then in the balance of ying and yang, interestingly enough. Here are some equations linked to this algorithm. Here's how some of them subtract out as an example;

Illustration/description 5.

12/12 $12-12 = 0$

12/34 $34-12=22$ breaks down to 4

12/56 $56-12=44$ breaks down to 8

12/78 $78-12=66$ breaks down to 12 then 3

12/9 $9 - 12=-3$

incidentally $-3 + 3= 0$

56/12 Twelve representing the Zodiac is 4.666666667.

In keeping with the algorithm layout, the five and six get its zodiac division in the sequence, numbers four, five and six, the six being repeating eight times. So the sympathetic resonance of numbers, shows this as eight after the seven, this demonstrates the close and seamless proximity of the planetary zodiac, in fact one and two is three beginning the number sequentially, the six relates to the nine being what seven and eight are added being a six, just upside down nine. If seamless, then this illustrates that five also relates to the female and the binary circle which is female and the six and the one for the male.

Chapter 6. Twelve represents the one in twelve

(3, 4, 5, 6, 7, 8) works from the principle of multiplying of numbers. Between the breakdown of the four and the twelve in the sequence in illustration 6. we have the number twenty-four, this fits in with twenty-four hours in the day. Base for square using this table;

$1 \times 2 = 2 \quad 2 \times 0 = 2 \quad 2 + 0 = 2$

$3 \times 4 = 12 \quad 1 \times 2 = 2 \quad 1 + 2 = 3$

$5 \times 6 = 30 \quad 3 \times 0 = 3 \quad 3 + 0 = 3$

$6 \times 4 = 24 \quad 2 \times 4 = 8 \quad 2 + 4 = 6$ Note; this is in reference to 24 hours day use.

$7 \times 8 = 56 \quad 5 \times 6 = 30 \quad 30 + 0 = 30 = 3$

$9 \times 0 = 0 \quad 9 \times 0 = 9 \quad 9 + 0 = 9 \quad = 9$

If you add all the end numbers, inclusive of the noted reference to 24 hour time, the resultant numbers being, $2 + 3 + 3 + 6 + 3 + 9 = 26$ the exact same number as the alphabet which the algorithm is based on! And minus the 24 hour reference which is the 6, it is 20 though inclusive of the 6 it is showing it is synonymous with the end result of both the alphabetical and numeracy characters.

With the illustrated example following, I used the synchronicity of numbers to blend in where;

$2 \times 2 = 4 \quad 4 \times 0 = 4 \quad 4 + 0 = 4$

$8 \times 8 = 64 \quad 64 \times 0 = 64 \quad 6 + 4 = 10 = 1$

$10 \times 0 = 10 \quad 1 \times 0 = 1 \quad 10 + 0 = 10 = 1$

With this you get to the remaining $4 + 1 + 1 = 6$. The synchronicity follows through into the 24 hour time.

Chapter 7. Key coding numeracy

The way in which the universe may operate according to the algorithm of alphabetic and numerical numbers, is to stretch out, much like a continuation of the algorithm, much like a chain of zero's. If space has area time has, width, height and breadth, then you could add a zero. Twenty-six would be, two-hundred-and sixty. So with the *illustration* 6;

2 6

2 6 0

2 6 0 0

2 6 0 0 0 add infinitum.

This would show expanse within a universal spacious field. This would inevitably be space using the same geometric layout as before, that which was described with the chapter on 'the cube' function. It could be likened to a double helix pattern in the DNA, like a strand molecule!

This molecule has hypothetically/metaphysically speaking a memory pattern in it as well. If this being the case, then it should arrive at it hypothetically speaking, more evolved formulation, thus the link in the chain. 26 may just as well be any variant of numbers, say 46. this could follow on with a chain as with our alphabet. The hypothesis being that if there was a more advance civilisation further on the time line, then this would be a higher variant of letter number construct. So to *illustration/description 7.*

4 6

4 6 0

4 6 0 0

4 6 0 0 0 ad infinitum again, completely random, though in context for this hypothesis.

The zero's, I'm surmising to be the catalyst or holding of information, much the like the embryonic stages of a cell before it splits and becomes one or more.

The base is the number and could be any number as example 26 our alphabet/ numerology algorithm, depending on evolutionary growth, The following zero's holds the potential to the surrounding universe from a metaphysical or Gnostic point of view.

A life cell at it earliest stages of growth, could be a civilisation/construct of 1 0 to the power of 100 as an example through to 100 to the power of 1000, as another. Please remember this is hypothetical.

When metaphysical activity crosses over into other realms of science and literature, hypothetical interpretation can be used to metaphysically calculate an inanimate object with an animate, physical one.

Chapter 8. Twenty-six in algorithm

Twenty-six represents the alphabet and it's algorithm in convergence with the numeral, or numbers. I've covered the interactive idea that the master numbers have a relevant place in the distribution of coding, that of one and one or a circle above or below another circle. This could be represented as two circles, one above the other, though in a vertical presentation, the eight too, has a close resemblance to the DNA helix. *Illustration 8A and another illustrative comparison of this shown in 8B with the binary comparisons 0 and 1.*

Illustration 8A

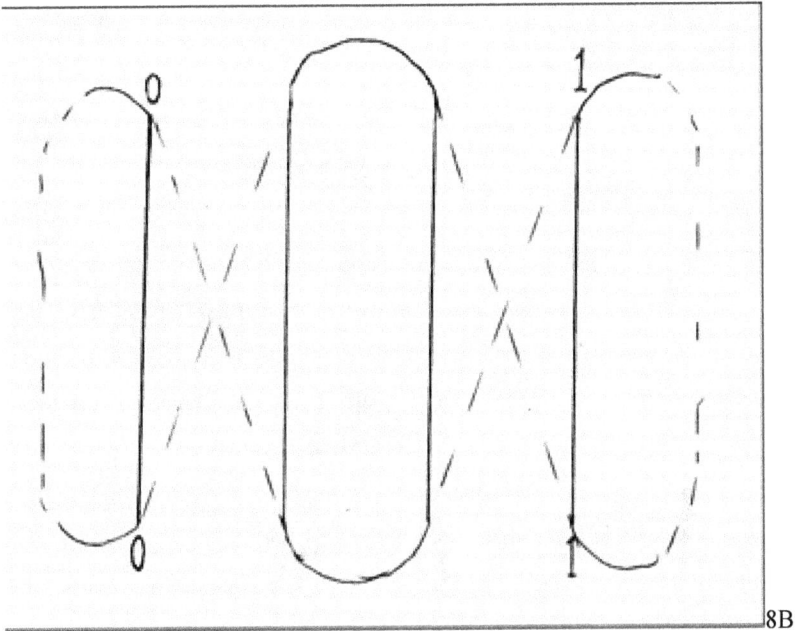

8B

Illustration 8B

0 and 1, notice there is a DNA helix associated with the form. Each link or strand helix, can represent an evolution. This can be configured into the key coding numeracy into the many evolving aspects. Illustration 8A illustrates this in that the DNA helix when looked at can represent the microcosm, that within that strand lies another of the evolvement of that strand and that the macrocosm represent the physical completion of that strand in the microcosm. As above, so below. The DNA connects to the numeracy of the numeral.

The idea also two and six could be in another representation, a parallel to a grid reference and taken into account with the illustration of this in the 'key coding' paragraph, whereas example; civilisation/construct of 10 to the power of

23

100 as an example through to 100 to the power of 1000, with as many zero's, one's and any of the numbers in between can be regarded and as another by in large, the difference in a civilisation may take on a uniformity as in the case of randomness or un-randomness of numbers as contributes to it civilisation and any wonder, that architecture has a distribution of shape and conformity as well.

Back to the grid structure and it has numerous boxes or cubes, each with their own reality. When I mention, 'their own reality'. I mean it as a paradigm. That isn't to say the paradigm doesn't have other numerous bodies or other paradigms interrelated with it, in the cube of space itself. It has been explained that the universe is crisscrossed with energy currents of different energies, this may have something to do with this phenomenon. As the illustration shows, the squaring of the circle is reminiscent of the factors pertaining to the cube and the pattern of it as a functional form, even an entity. These would be symbolic of the resonance of the patterns, usually in the most basic structural form as is the atom.

The link with universal principle colluding with a path steeped in accumulated and intercepted function, is an apparent manifestation of will. Through this will, the result ensues.

Politics can be the initial spin offs of resultant discovery of fields of the architecture, (the method used for anything from mathematical calculation to building the Sistine Chapel, mapping (which is the consistency of following a path, where we have been, whether marking off the peg for Sistine Chapel to the stone markers of trekkers following where they may have left off in an underground cave.) Flight , (being the function of the idea and reality of flight such as flight throughout our time, mythological Vimana's to technological flight.)and technology, (the technology that has been present throughout the ages, mythological and the consequence of technology in the present time from the introduction of the Industrial age, to the invention from everything from computers to aeroplanes.)

The construct of an atom could be the link between crystalline and biological (animal species) will inevitably be linked to the closest representing structure, a dodecahedron shape of the universe. Thus shaped due to the subtleties of the planetary bodies as opposed to those that dwell on them, the lattice work and nebulae of the universe itself. It is more like a spore, this dodecahedron, a free floating crystal, biological spore. But how could this be? It is a supposed truth, one based on theory rather than true evidence, that if the universe can evolve within the spore (organic) like dodecahedron, then it may not be a stretch to believe otherwise of the idea of a spore (organic) atom. This crystalline growth of the universe growing outward from within and without outward, far are not both crystalline and biological forces at work on air earth water and fire elements? The lines of force in the universe which unlike wolf's law on the explanation of loading of bone the Human or Animal will adapt to the pressures of the bone loading, the biological evidence of stressors on the growth of bone or changing of crystalline nature of it within the body of Humans and Animals and could in this case be true for the same lines and energy of force found within the universe and effect structural growths in the outer macrocosm rather than the inner microcosm. Found within this is the apparent theories of our small universe found within the structure of the universe itself though on a smaller scale. I bring this up with the value of number in Key coding numeracy Chapter7. We are living in a crystal which could have the characteristics of organic and inorganic structure.

*Illustration 9.*of a dodecahedron. The concentric pattern in it is followed by a use of a further delineation and exemplification of cube dimension and space, it also fits in with the illustration 4F and 4 G with squaring of the circle and with the .more so however with 4F in that the triangle has the angle or crystal form, indeed in nature nothing is truly square, when looked at from the microcosm, the macrocosm can show the manifestation of angle, though this is alternating at the changes in these perceptions, as indeed the viewing from a microscope of these things can also testify to.

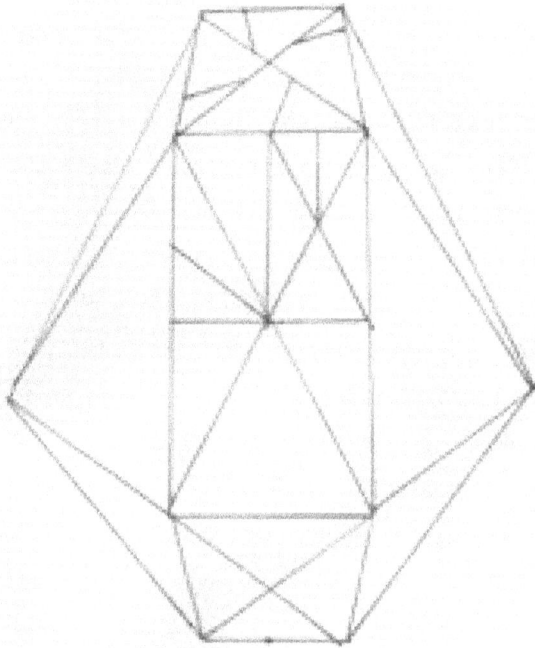

Illustration 9 Dodecahedron (type)

Technology is based on nature, the nature that is related to God. The God factor is found in natures fauna and flora and is related to the oceans and all that is in them. They are related to finer higher tones of the planet where humanity abides, the lower tones are in the rocks and associated metals in the Earth's crust. Up until it becomes recycled back to un-associated tone with that which is higher and on the surface of the Earth and found above in lower in degrees in between and throughout associated. The terminology, that which is above is also below as said by. Hermes Trismegistos.

The higher vibrations, mirror the lower vibrations. The equal and opposite reaction that drives the planet we live on and her surrounding cosmos. There is an apparent dual purpose in the foundations of the Earth, that must mirror all that surrounds the Earth itself. Many geomantic points on the Earth are in relationship with the numeral/alphabetic algorithm as is the cube of space dimension. A display of this could be described as the squaring of a circle is shown in the next illustration. As in *Illustration 10.*

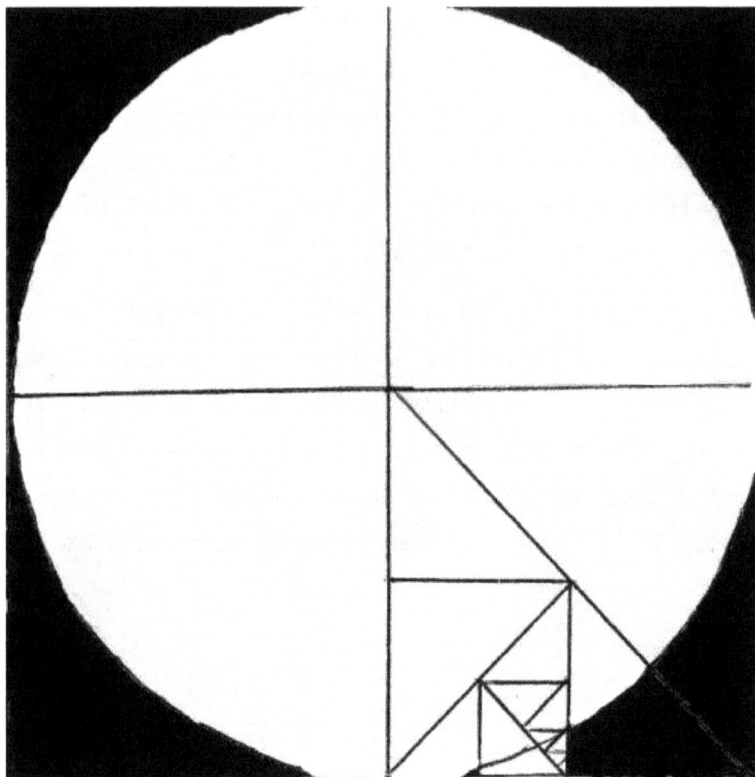

Illustration 10 A field of further delineation into another cube and space to a final result that's am orphic cresting the flow of rounded shape, although it's structure before repeating the cycle again to change slightly, a theory of alternative reality.

Once a field is worked out, mapped or projected onto a diagram, another set of repetitive structures of the same semblance though in a different space will come up, such as what is found in the above illustration. Not only a circle, now that it has been graphed in diagram, we'll have plotted into it, the squaring, suffice the squaring of the circle and the triangle. This method help with the next dimension of field, that which is off to the right of the *illustration.10.*

In the next description, the display is brought forward magnified, so that dimension is better seen and understood. *Illustration 11A.*

Illustration 11 A

11B

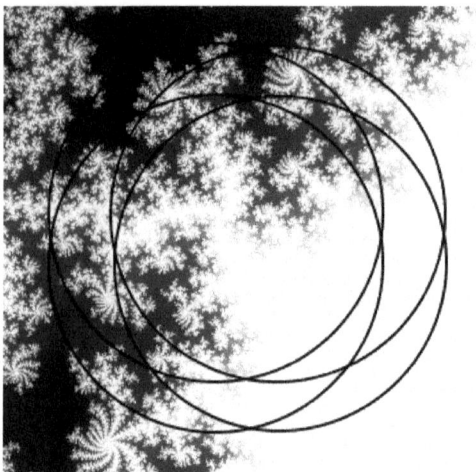

11C

Illustration 11C shows that the atom could be mapped as on the surface layer with different reactive traits both positive and negative and also neutral, there is a disassociation with the far right circle (atom or circle of life) the 'Mandelbrot' is the perfect example of the universe of infinite structure , which represents the element of 'form' or conformity of change associated with the atom itself (in theory)and the ever present possibility of the theory of random occurrence within a greater whole. It is a randomness of shape and function over-all, for a solid and is the dissociative element of form, over-lapping the same form, which in theory creates a new element in dimension. The *illustration of 11A, 11B and 11C* shows with the top circle dotted points 11A demonstrates simple structure one and two dimensional, these could be the behaviour exhibited by atoms as of the microcosm which could tie in with illustrations 8A and 8B explaining microcosm and macrocosm behaviour and has a further extendable element to the description of the further delineation of the cube of space again using the squaring of the circle. It can be seen here, in the use as in illustration 4, *4A.* Shows the perpendicular lines of structural core, now shown in half value to the squaring of the circle, which is derived from the cube and therefore the square is a singular reflection of the cube, the circle of the globe and the triangle of the pyramid, all inter-related to each other, finding all their values in numeral and alphabet terms.

Chapter 9. The spectrum and tone

The colour spectrum of space and the cube is demonstrated primarily by the separation of the primary colours of the spectrum, for example red, blue, yellow. White and black fit into a differing correspondence to the order of these primary colour that in turn can separate further into colouration through degrees of manipulation, such as purple green, crimson brown and so-on. It is dark matter that fits into the mystery with light matter, just in dualistic correspondence both with different applications on the dualistic scheme of thing.

Finally 'tone' has a capacity to work with the correspondence of colours, sometimes keeping these explanations short helps. The tone and colour correspondence has the application of which there are at least three, 'static' colour or tone in a neutral position and/or moving energy will produce a differing result. The dark matter is such because of depth of darker correspondences to the light energy. Some depths are greater than others and create in the deeper veil of the universal fabric, a rift or overflow of energy held together by a sympathetic factor of universal abundance of such matter that it will because of space allocation tone and colour will correspond to that 'frequency'.

Chapter 10 Space bodies and the Mandela

The chapter here I've added after further recollection of researched material. The early period nature religions had ideas and furtherance of ideas regarding the space bodies, the planetary material and why it was established. The universe, the zodiac and planetary influences, indeed anything regarding the planets could be attributed to these early religions and civilizations. The Gnostics explained that the planetary bodies were the work of entity spirit energies, naturally migrating to these regions. As the universe expands and contracts and has always been this way and after which, the created Mandela's express a subconscious delivery of visual information, opinion can widely vary on the nature of Mandela's but it would appear that they centre around the deep religious, esoteric areas of information. The explanation of this, I believe can be the Archon and its, Archaic deriving can be argued. The deriving of a space-body has to be explained and if an energy from a completely different locality in the universe is travailing from another galaxy and effecting another galaxy in the process and then it is no question of it being linked, this can be the beginning of an Archon, though not always. How can that be explained now with the region we call the here and now. It is a mystery, though the energy which put itself there in the first place is not being fully understood and for that matter not really taken any interest of at all, placing itself there if it's in or on two regions at the same time. Here's the stretched answer. A Mandela is energy, the budda image can be a very good explanative example of this quantifiable result of the classic Mandela image as seen in the Mandelbrot, continuance pattern as seen in *Illustrations for 11B and 11C* , so too are we. The universe as explained by some researches is filled with beings or entities of different varying consistencies of ether atmosphere's or energy felt and perhaps not felt. Quantifiable reasoning is varied and possibly erratic based on the scope of what is researched. Here's the stretched answer. A Mandela is energy, the Buddha image can be a very good explanation example of this quantifiable result of the classic Mandela image, so too are we. The universe as explained by some researches is filled with beings or entities of different varying consistencies of ether atmosphere's or energy felt and perhaps not felt. Quantifiable reasoning is varied and possibly erratic based on the scope of what is researched.

What can be attributed to the ether plain, the microcosm and macrocosm is the space bodies, just as there is an outer space, there can be an inner space with the same causation of Mandela taking place? Is the energy that has been explored in the more local areas such as science and alchemy probed this as well. A perpetual pattern of the same and not the same out of phase and in phase at the same time gradually moulding and morphing into the nature we possess all around us. An over lapping of this microcosm, macrocosm alchemy demonstrates the extreme pole of opposites in ourselves and others, from the constructs of darkness and light, all with the illusion within and without, within the space bodies and without the space bodies.

The Archon and Nature demonstrates that the whole field of the universe is geared toward entity establishment, these range from physical fixed bodies, for example a planet, a more subtle field, a nebula and to finally invisible fields of matter but may be seen through a microscope, infer-red and other spectrum analysis. The fields consist of subtle life energies that may or may not be seen, when again dealing with the macrocosm and microcosm in analysis, then the whole idea of the inner terrestrial universes found at the microscopic level of for an example, that of bacteria and smaller, that within these fields, that it may introduce a conclusive look at details dealing with the spectrum of all existences of life through the Mandela and in effect how things better function.

The Mandela has with it the same scope as the grid of reference in that it can be a determinant of distance structure and shape, thought the whole idea of it fits in with the longitude and latitude of determined points. Archon energy and vibration is much the same as ours, though it would be operating at quite possibly a higher frequency, due to the idea or the fact that it is on another plain of existence or dimension. When toying with the idea, Archon entities are represented through this continuum of a logic that is part truth and part non-truth when dealing with it on the physical plain of existence.

Reference to plains and sub-plains, which are all very prevalent on the space and grid rooms and for example; space area atmospheres, can explain the different qualities and frequencies of a message. The message that the Archons want us to hear, would be completely irrational to the normal mind. The mind is a transmitter and receiver and this determines whether we get the message or not but to the Archon energies, depending where they operate from in terms of space and time, will be doing for example work on higher or lower area vibratory rate in the aura and chakras regions of the body are all of one vibration angled in on the keying in of light or using the perspective of 0 and 1 as explained in the first chapter of my book, the base e grouping of 0 and 1 binary, there is the sound tone spectrum and frequency of the colour spectrum, within would produce it's unique energy according to the Archon receiving station outside of itself and to us as Human beings. When received the interpretation according the field will be decrypted and interpreted by the Human mind. The Archon may have a completely myopic and benevolent means of sending and packaging the message. Quantum theory has many explanations, though when Archon energy is explored we can find the answer different to what we possibly expect, expectation of a result may or may not give an answer that is workable. As with the atom explanation in illustration 11C, with the theoretical change of an atom, just as a cell in the changing of this to a human embryo, so too is the change prevalent though can be varied regarding Archon energy. So too there may be a macrocosm and microcosm at work with them. Lastly illustration 12, shows the infinite negative and positive and which as an energy, physical or non physical on whatever frequency from this illustration is what radiates out is the prolonging condition of infinity which moves out at a quantum level from any particular circular or point of radiating direction affecting and changing whatever it comes into association with and with this, the equal opposite effect with illustration 11C. This demonstrates the physical universe, the synergy of centrifugal energy, this also shows a completed circle, also representative of a helical pattern. *Illustration 12.*

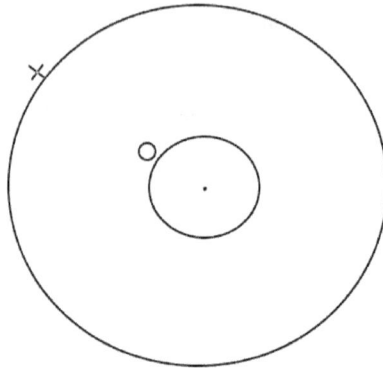

Illustration 12, the synergy of centrifugal energy.

In the 'Nag Hammadi', we have the portrayal of the Biblical character of Eve, who corresponded with the interested Archon, through a means of keeping apart from it physically, this is when it had in mind, a coupling of its energy with hers. She created another body apart from herself to couple to the Archon as the story goes and its wishes for coupling. The premise of the story is, the Archon is an energy, though the frequency is apart from the Human races and the nature of these things are found in texts going back to Qumran and the Dead sea scrolls.

Chapter 11.Space bodies and the Mandela conclusion

1.What result from all the information presented, is co-existent on the factor of frequency, the application, or reverberation of response to the space of any one entity or entities, whether it be an atom (female, hermaphrodite or male, organic or inorganic).

2.It's application to the algorithm in a structural and/or a morphed form, this could be an emotion to the building of an application via creation for example; sacred geometry and within that produce architecture, design, technology and mapping and language for the written form, alphabetic numerical and logarithm for the application of these in the long term sense and that of the Akashic record of knowledge of the universe. In the future as science and application of technology architecture and mapping prevail, the continuance of religious and philosophic ground will enmesh further, whether it is rough or smooth in its production this can be determined with time. Love in the correspondence as an emotion has a part to the use of application of the above mentioned to bind and make sense of it all.

3. Quantum level frequencies or tones are found throughout the twenty six letters of the alphabet to effect change at the quantum level (this can manifest in the different paradigms of existence). From one quantum whole as is binary zero's and one's, the female and the male to effect the structure as illustrated and described in illustration 11C and 12 to produce in the final conclusion the child or hermaphrodite. The three is a structure very natural to the universal principle. 3 x 3 is 9. 9 is related to the unfolding of the golden mean principle and is found in every geometric, geomantic and geographical and at an individual level, every man, woman and child is '3' and every multiplication of it i.e.; 3 x 3 can be looked at as 2 x 2 for the male and female principle is '4' double that, it's '8' for the infinite unfolding as well, symbolically. Now there is with the 3 x 3, the '9' and demonstrates individual, is microcosm and can be macrocosm though more suited to the former, the geographical, etc or 'geo'

states are representative of the macrocosm, again can have instances of the microcosm. Stages cumulate to the end of a cycle and in the numbers, demonstrates this. The square or one, fuses with the circle and in the end, the principle has a 'square effect' as I call it, so for i.e.; '6', that can represent the zero and a one on inspection of that character number if we get into the detail of the symbol of that character number '6' and although has the same but opposite symbol of the '9'. When looked at on a continual basis, every number is the same in subtlety though the principle of nature found in the squaring of the circle is found mainly in the zero and the one and is as quantified as the principled frequency or tone as explained earlier in this chapter. From the quantum perspective, it's understood.

The tie in with the universe when looked at the principle of number, letter and tone/ frequency, the orbit for the twenty six tonal sounds are radiated out until critical mass changes this completely to effect the next tonal frequency. If in theory, another tonal frequency altogether, sort of comparing capacity with frequency to effect for an example, another form, or change, much like what was outlined in illustration 11C to give another set of symbol energy.

Chapter 12.Development of theory

The conformity of number, the disintegration of number is the same as the letter. The letter emulates the pattern and consistencies or not of the letter, to encode to whatever structure or form is permitted within the letter and number. Structure being smaller or larger, for example, the atom, does the simile of the number of letter vice versa fit into the shell of the atom space, the fullness of the atom space or to the neuron's within it the size of footballs compared to the entirety of the atom. This will influence density and if smaller as the football size, will not have the same strength as the fullness of the atom. Yet with that said, the energy is different in the smaller as it will have an empathic response to the smaller mirrored universe it linked to or /and allocated to, whereas the atom fullness can have the same link as the atom , it's energy, even though is familiar to the neuron, will have a different energy and mirroring of the closer physical reality we are all used to. Another way to put this, is the neuron is mirroring or closer to another dimension or reality of space/time.

With all this in mind, the Pythagorean system is related to the whole aspect of shape and geometry, the structure and the sequence of the formulas are part of the structure of the universe. And the conclusiveness of this is that the number and letter system is formulated.

The displaced/ fraction type squaring of the circle letters and numbers, fit into the rarer specifications though no less important structure, usually corners, not width, heights or depths. Corners place more significance on the shape, structure and fixed settings; ie: example, the optimum strength for the given structure. Form as opposed to total non-form.

Sovereign application of the cube as in the significance of the shape to share, reduce, deduce and render of the 'letter and number', literally is the base and the basis of the 'pyramid'. Geometric shapes and architecture has it's root in the shape of the square and the cube, the variance of universal structure. As I've outlined before, the basic underlying basis for matter is, mapping, technology and architecture, structured into a cube. 1.Mapping is fluid like our blood, metaphysically, it's our thoughts and what drives thought, is blood. 2.Technology is akin to our body, for electrical current, electricity, the physical driving force. 3.Architecture, the complete physical application of the physical movement of a body, the body.

Now all of this is not only through lattice structure grid work as a continuance of the square but the vibration energies and light that give it support to the rest of the structure to form the cube. The theories of structure concerning dark space and (controversial) as it is, give credence of lattice, for example; when a pattern in a quartz crystal, shows a crystal lattice and when you turn it to the light, it beams back throughout its own structure, invariably amplifies through the light. It's a receiver and deliverer of light. The uniform rather perfect lattice pattern that shows up as light. The same is for dark matter though to extend in further into the understanding of this, an example; when you see a fence and then watch it from the sides, it gives you just the side view, the most minimum. If that fence was instead crystal inside, it would be considerably darker from that side view because of density of shadow, this darker than from the front view.

Some reality or dimension however, it will be considerably change because of extrapolation and not jusr of a similarity of the letter and number sequence due to subtle changes in the time/space process, this process by the time it's been through formulas by the use of space, then it then is a vibration.

Letter and number is used in sequence or as a consequence, natural and artificial fixtures of information in the process of making physical matter to (letter and number) gives the required order necessary for the formulas to be even accepted and passed and/or not through natural or un-natural processes. Some of these sequence of information can be floating due to radio waves and can be generally sent in the affirmative in differing clarities due to the responses accepted and with all this in mind, the Pythagorean system is related to the whole aspect of shape and geometry, the structure and the sequence of the formulas are part of the structure of the universe. And the conclusiveness of this is that the number and letter system is formulated through the processes that formulated Pythagorean theorem and/or process.

The displaced/ fraction type squaring of the circle letters and numbers, fit into the rarer specifications though no less important structure, usually corners, not width, heights or depths. Corners place more significance on the shape, structure and fixed settings; i.e.; example, the optimum strength for the given structure. Form as opposed to total non-form. These geometric shapes and architecture has it's root, the variance of universal structure. As I've outlined before, the basic underlying basis for matter is, mapping, technology and architecture. Mapping is fluid like our blood, metaphysically, it's our thoughts and what drives thought, is blood. Technology is akin to our body, for electrical current, electricity, the physical driving force. And architecture, the complete physical application of the physical movement of the body.

Now all of this is not only through lattice structure grid work but the vibration energies and light that give it support. The theories of structure concerning dark space and (controversial) as it is, give credence of lattice, for example; when a pattern in a quartz crystal, shows a crystal lattice and when you turn it to the light, it beams back throughout its own structure invariably amplifies through the light! It's a receiver and deliverer of light. The uniform rather perfect lattice pattern that shows up as light. The same is for dark matter though to extend in further into the understanding of this, an example; when you see a fence and then watch it from the sides, it gives you just the side view, the most minimum. If that fence was instead crystal inside, it would be considerably darker from that side view because of density of shadow, this darker than from the front view.

The rings of life are comparative of the sub-structure in the universal patterning, they are the beginning and end in a continuance pattern/destiny.

I have the semblance of them structure and color coded and this is more rough and abstract in reality, it is as if the way to look at this and indeed all energy is that it is energy in motion, elusive yet definitive in application, structured and yet comparatively not. The rings are fronted on logic, They reveal not only the fact we live from transmigration/ transmute our life to the next, show also that this color coding is holographic and is synchronous in universe oneness and diversity harmony and/or disharmony and transmuted into that, what its former apparent was though now transformed though transmutation making a change into a new dimensional force and/or change from the perspective of the universe of a new life. Reality is by far a superior standard to go by, however as I extrapolate this further a change precipitates and/or changes whether subtly and/or majorly and I've reiterated, it is not easy to grasp and take like a physical object, although science has these determinants based on of course advanced enough applications i.e.; Technology, Architecture and Mapping which is in itself the schematic for understanding. The next *illustration 13* shows the singular life rings.

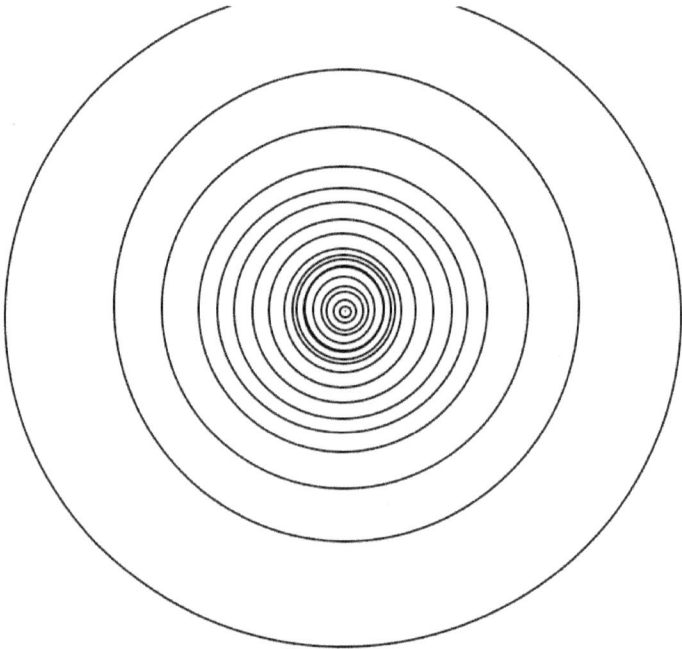

Illustration 13. Life rings of transmigration, of one-singular

All these things of the structure have something that reveals that we live in/on transmigrate and transmuting properties i.e.; Earth=Yellow Square, Air = Blue

Circle, Fire=Red Triangle and Water =Silver or in this case Brown (Planetary phases, Moon, Luna) as in the color coding in Illustration 13A.

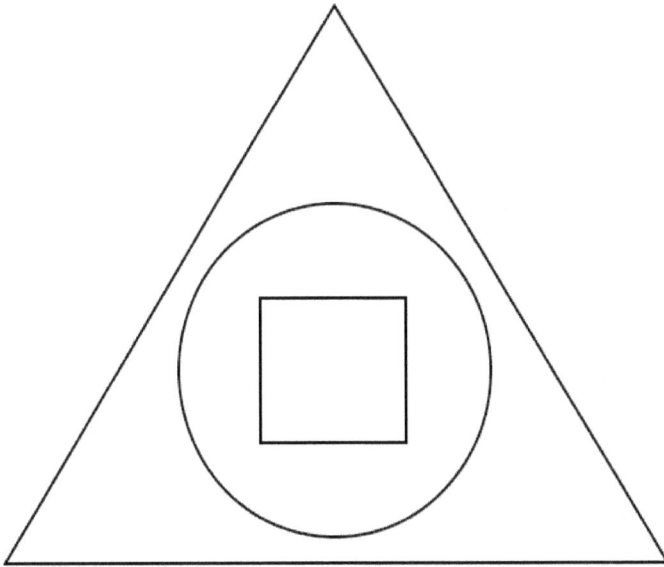

This is illustration 13A.

The atom field of the theorem would require structure, i.e. as mentioned above with Circle, Square and Triangle and I must include the combination Silver, Brown color coding referenced and this would apply to the transmigration-multi-collective-duality and its color coded component example. So the one in the Illustration below demonstrates reference of shapes within DNA. This is the event horizon (the triangle for example of the shape of the 'Circle and Square' layered structure, all shape make for a better geometric pattern, shape and structure. Take one of these systems away and it will most definitely be more voided of a coherence whether in the anyway shape or form.

Below, *Illustration 13B.*

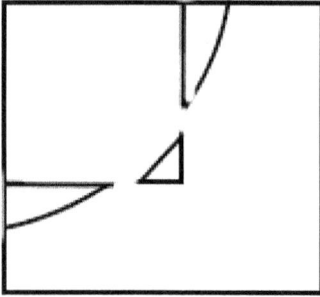

Life rings of transmigration, of one-singular (atom theorem as explained)could have a predominance of triangle behavior.

The next *illustration 14.* shows the structure of the, life rings of transmigration-multi-collective-duality, and in illustration 15. we have the color coding in the pre-supposed holographic of the life rings of the same, transmigration and can be transmuted into what it was in the previous illustration and is a hypothetical theory of life cycle jumps and with the next, we are now describing an inverse/ obverse reality that expands its space bounds, taking on that of a multi-faceted reality structure that still takes on its singular structure namely this physical reality and not at the level of the atom! The atom is in fact what has the ability to morph and to change, again found in technology, architecture and mapping i.e.; mapping sequences derived from the interactive nature of these. This illustration, *Life rings of transmigration-multi-collective-duality* like the Mandelbrot, life, rose that's has split realities can be a good example of this.

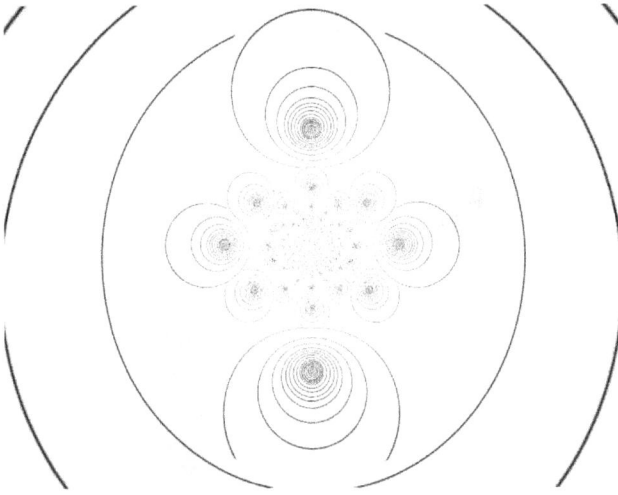

Illustration 14.Life rings of transmigration-multi-collective-duality

Now peaking into a new dimensional reality perhaps made into an artificial perception and/or perspective to understand this application studies involving many hours of work can bring generally a new perspective of the magnitude of what is created/made. It makes a conical shape much like that of a chakra on the body and the atom has a space empathy that would be theoretically linked to life rings much like in the marking of growth rings and as they delineate or spiral up or down and as with *illustration 12.* negative and positive forces would create the stabilizing effects of this harmonizing integrity and is prevalent in *illustration 13A.*

Illustration 15. Another filled in Life rings of transmigration-multi-collective-duality

Knowledge has its limitations if its the wrong part or parts, any disproportionate and /or misrepresented then you will not get the correct formula but a more extrapolated path of resistive and unreceptive energies

The magic of squaring the circle.

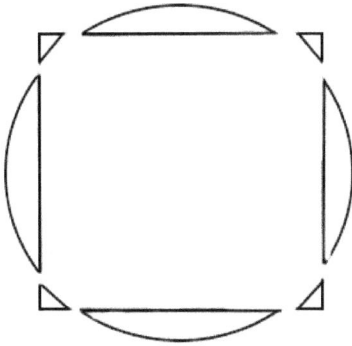

As illustration 16. shows the squaring the circle is illustration 14.

The pyramid of correspondence shows four corresponding parts or areas of the pyramid to be 4 equilateral triangles. These triangles superimposed on the rest of the assembly denote the apex of the summit of the pyramid but also the cube or more appropriately the cube is space oriented, inverted and obverse in a cube like structure brings together a fascinating pattern.

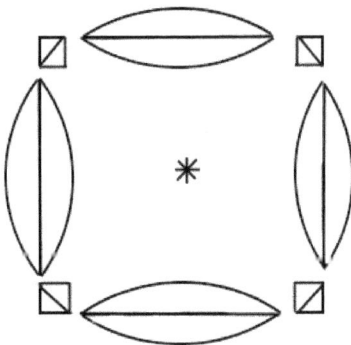

As shown above, *Illustration 16A. The 4 equilateral, above and below structure of the cube. can be the base of the pyramid.*

In the centre we have a particle generator/generator, usinga guiding direction of existance, depending on the material, we have a line generator of structural significance.

The centre of the pyramid is omni-present, meaning its everywhere on every pointwithin the centre. Its on top, below, middle, side underground, which ever way you look ast it. The cross over section is of basically squaring the circle has a limited though varied way of existance.

The above and below sequence is depicted in the pyramid and is amazingly complex but has substantial work only in balanced consisting then appropriately as a cube as far as above and below. The sequence having the form and function to enhance what is derived from the sequence, As above is as below, yet surprisingly on the space sense of above and below, air and earth, the air verses air, earth verses earth is apparent then the conclusion to this is apparent as well then the conclusion to this.

Omni direction as if being repelled by the source of it, the centre. The omni directions is fixed on the floor of the pyramid at the centre.

And as it is what we only a personal and singular way in this configuration, paradigm and the flux there off. It's not as clear cut as a paradigm, it can be a reset of sequences and input put forth every day by the inner mind or sub consciousness. Deliberate tampering, whether from within and or without meaning outside of ourselves, brings this fourth.

Chapter 13. The Truths

In relation to the truths of ; Letter and number correspondence' and the corresponding negative space factors or dark space area's is related to a fundamental vision of the lack of dimensional for example: One and two dimensional. The first dimension is of corresponding space characteristics that are dependent on their functionality of illusion and/or of the Archon, the hierarchy of letter and number or vibration. Their related space dimension that's used within the context of it/their illusive divisiveness. This may seem of the track for example, on track to be off track with the hierarchy deciding on as from the starting point of perspectives of the usage of the function of illusion within the confines of space.

They, it, what therefore is represented are liken on a subtle level as to emissaries or diplomats within the letter and number landscape. The understanding of the environment and co-related factors of the first dimension or the starting point of negative space, then successively positive space to first dimension perspectives, then succeeding second, third, fourth and so-on dimension supporting the first dimensional perspective then sub-dividing lattices of dimension for example: a second dimension supporting a 1st, 2nd, 3rd and 4th dimension opposite, infinitely.

The Archons, the elemental type spirits intercede. Physical to air properties and successive dark to light fit their make-up, they occupy lower and higher levels within the lattices and dimensions described and further, depending on densities of existence within their scope of function. The illusiveness may be found in the following examples: anomalies and in function that did not appear apparent at the time within the scope of the personal mind of the perception of the viewing participant and an Archon or inventor and/or maintenance engineer of one of the many examples in the category, they can then be found in the many lingual dialogues throughout historical discoveries, the outcome of which generally is found in the field of Archon invention.

DNA and the language and vibration make for the changes that make up the over-all time sequence to everything in the body, the physical function that delivers all corresponding levels and attempts through random, causative and purposeful change within the body and involves sound and light, which could be in sequence or out of sequence. I mention attempts because the source of natural function is non-artificial but has a definite nature's way about it. It could be circular in function, growing in spiral, hexagonal, straight, wavy even in seeming un-natural ways, though always organic as demonstrated by the life rings of incarnation.

Archons, and the definition of the them is an entity that resembles the life forms around them. They have a dependence on the life organic around them, it is as such when they appear in Delos, Greece in their mythology and they are misconstrued as the intentional creation. Their creation depends on belief. This also mirrors what an entity accepts. By definition it could also mean a deception. The deception is rather sequenced, it makes belief in the most basic form, which is why the childhood drama's are at least considered to be fairy tales that have some reality though on a subconscious level. When all of this is considered, then it is role played, this role play can bring circumstances with the childhood and into adulthood that can influence all that remains with outer circumstances i.e.: stage performances, books and testimonies of these influences wrought by the Archon influence.

The topics covered generally suggest what could eventuate with the sequence of events Archon or otherwise seemingly related and lining up and generating a tipping point fusion event. That will be eventuated as a prediction to this and the struggle for this sequence of events also brings light, to what is

50

demonstrated in the whole aspect of the approach between light and dark, they reflect opposites though between one another.

The number and letter continues in the past and it expands into an infinite time, it can only find another reduced point of simplification, sort of like a code formula but on a larger scale. Perhaps an opening portal/door to another similar existance which may be found another sort of density, the universal continuance, and if the continuance is further extrapolated further out, it may get differences of the vibrational factors and calibrations no matter how subtle, just different. Any face of the mapping (geological and/or eons of space inclusion. Architectural, (with formula's, templates of what was brought and again extrapolation). Technological, just on the premise that technology could have sent the extrapolation through.

Space is a void though still needs to be filled. Filling it with form and there are many distribution channels for this, once of course is the message that is sent and the structures of space that are filling up more denser patterns, formula and so on, the sequences can attach more rapidly, sort of like a magnet.

The diagnosis and break down of this is: 1.Firstly, letter and number, with the symbol, word and ratio of geometry in its greater meticulous arrangements. 2.What is in this table of representations are have been in reality formed, made and demonstrated in different ways over time and space and this history gives us the formula/s, the equation/s and symbolism/s and even mythology of meaning but if we could get into the more intricate, mediocre, semblance/s in whatever vibratory means they send out the signal (projection of what the form solid is and the variations of these behind the quantum of which it is, of which science has made the discoveries on some at quantum level. Take again the quantum to what end it's become and by perception further enhance the effects of the previous formula/s, the equation/s and symbolism/s and solidification of the composition of the variances of vibration, we would have the continued mutation of the solid being structural, technological and biological. The measurements, insights and readings of these letter and number representations from the smallest atom reference crossing from one reality to another with each atom representing a letter or number of identification to the greatest solid capacitance of physical structure, i.e.; a cathedral or a piece of technology, tipping into the reality that is known through to the unknown. All of these have their basis of the squaring of the circle both singular and doubled effects are provided here, including the letter and number doubled as well.

In relation finally to Alchemy, the letter and number hypothesis is complete through processes as best described via Geber as he describes the states as linked to forces of planetary influence, the link be that indicative of the following terms in conclusion to the processes; Calcinations, Sublimation and fixation, Coagulation and Fusion, Dissention and Calcinations , Coagulation (as in a water bath). and finally unrelated dissolution, this then represents everything.

Disintegration , reintegration and the topic at hand will explain the process of the graph/algorithm that intercedes with line and letter and number. The process is an end result wherefore the code of DNA is disseminated. In order to read it, the process is dissolution like that in alchemy is wrought into it there is a circle, there the square is equated into it, so is a triangle, the shape of influence is also a sphere, such as the sun (star), Moon (planet) and via the moon there is eclipse, and this is illustrated by this phenomenon and a partially eclipse of any planet, leading to a Luna like crescent, this we now have circle, square, triangle and crescent which represents a cycle of change in the DNA sequence. Such being as a cycle of adjustment to whatever condition persists within the body, so too the universe and the inversion is the world above, the opposite obverse, the world above and through the sequence of letter and number correspondence, is infinitely interchangeable. As shown below, *Illustration 17 with key.*

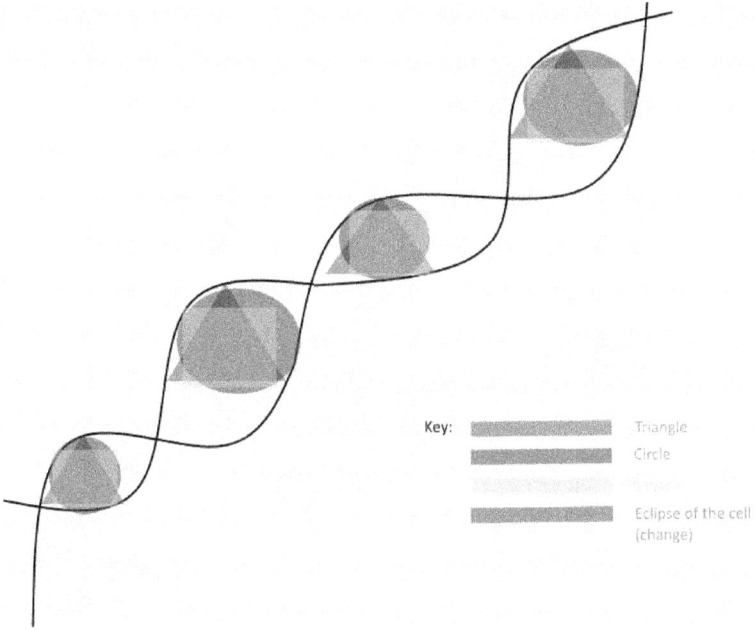

Key:
- Triangle
- Circle
- Eclipse of the cell (change)

Illustration 17 with key.

Reintegration and the destroyed object or form can then be referring to duplicating the aforementioned item and to the building up of the structure that has been built, wraught into form of DNA. Yet is presented as a cube of space.

DNA by product and consequently the form is made complete, it technically becomes a product (of universal standards).

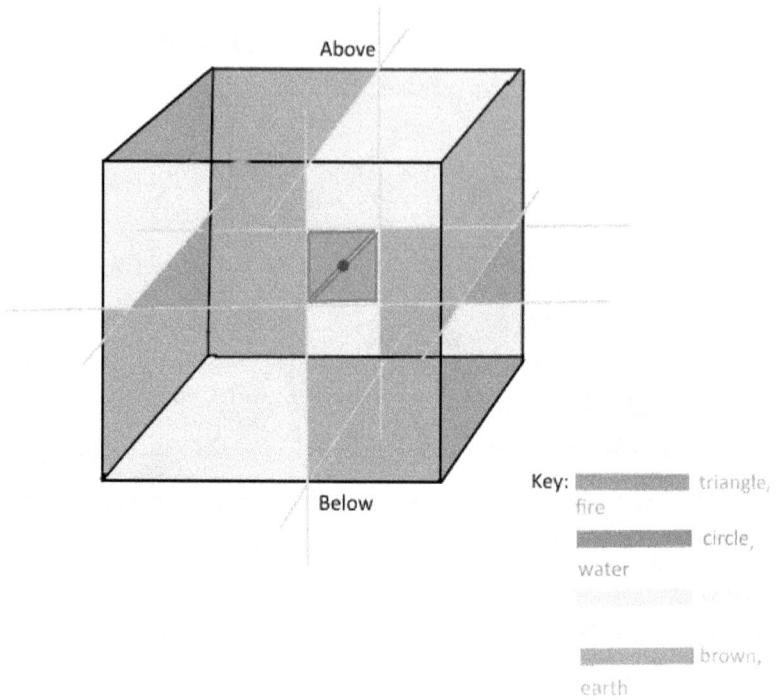

Illustration 18 with key.

The above and below sequence is depicted in the pyramid and is complex but has balance and consistency then appropriately shown as a cube, as far as above and below is concerned and the black hole represented as the color blue for water. The sequence having a form and development to enhance further what is derived from the square. As above is as below, yet surprisingly on the space sense; fire, water, air and earth. The elements compel one another and so the conclusion to this is the, direction as if being repelled by the source of it, 'the centre' being water, the cell, the atom. The directions presented are fixed on the floor, roof and sides of the pyramid at the centre. Another version of this is the split rose, showing the central hole in the centre, the triangle and using an alternative to this, in this case a trapezoid and the finishing square. Basically this is still a triangle although it has been capped off at the end so that it represents a four sided square, it has F,P and Z master number reference as well as mentioned in chapter 1and *illustration 1*. I mention this as the illustration begins with the binary 1 and 0 and the end of it 5 and 6, they add as follows;

$1 + 0 = 1$ and $5 + 6 = 11$ the break down is 1011.

All references pertaining to the allotting sequence of information contained in schematic graph/algorithm in stipulations and influences thereof. *Illustration 18 below.*

Illustration 18A. The split rose.

Chapter 14. Conclusion

Like the content of this book, there is reason in this final chapter to describe things in the symbolic, formulaic descriptions, reflecting and referring back to the source of the content and its evolvement and some of these areas of information were common knowledge, however my research has expanded or extrapolation on this information, largely dealing with the 'Technology', 'Architecture' and less widely known field, 'Mapping'. Mapping is an am orphic field that changes and interchanges in a graph ting way. I say 'way because it's sometimes manufactured that way. This can lead to what I believe to be a less natural, although more manufactured 'way'.

Shape and structure, like pattern and vibration is a wide field and this is what 'letter and number' corresponds to. Correspondence is communication and that's the way I manufacture what a formula is, or many formula's. If a claim to letters and numbers has been predetermined, then this then is a formula.

Cause and effect leads to cause and effect, though, like the dimension reflecting from one parallel universe to another is the truth, this merges then into 'other', not dimension but Archon. To me Archon is the reference of language and over time language has emerge as it has I believe, through the subtleties of the that energy, the appearance or non appearance thereof.

The End.

~~~~

Thank you for reading my book, if you want further information and updates, write to the author at the following address; Nicholas Betar, c/o Australia Fair post office, Australia Fair, Qld 4215 Australia.

####

www.ingramcontent.com/pod-product-compliance
Lightning Source LLC
Chambersburg PA
CBHW071737020426
42331CB00008B/2072